A Glossa Plastics in 5 Languages

CW00351902

English
Deutsch
Français
Español
Italiano

Edited by W. Glenz
in collaboration with
H. Guyot
R. Marchelli
M. Santolaria Mur

Carl Hanser Verlag München

ISBN 3-446-16226-7

© 1990 Carl Hanser Verlag München Wien
Druck und Bindung: Druckerei Sommer GmbH, Feuchtwangen
Printed in Germany

Preface

With this little glossary of plastics terminology we hope to breach the language barriers within the international plastics community.

In trying to serve plastics engineers in different countries, the idea was born to compile a glossary of plastics terminology in five languages. Although we are aware of the problem involved in such a word-to-word-translation of terms without further explanation or interpretation, it was nevertheless felt that such a "vademecum" would be useful for the plastics technologist.

We would like to invite users to submit their comments and suggestions for improvements to the editor [*]). This will enable us to update and improve the terms and translations as required for an up-to-date reference book.

We would like to express our thanks to Svenja Knauf and Klaus Knobloch who assisted in the preparations for printing of this glossary.

Darmstadt, Germany   The Editors
October, 1990

---

[*] W. Glenz, c/o Kunststoffe, Marburgerstrasse 13 D-6100 Darmstadt, Germany

Nr.	English	Deutsch
1	**ablative**	ablativ
2	**abrasion**	Abrieb *m*
3	**abrasive**	abrasiv
4	**accelerator**	Beschleuniger *m*
5	**acrylic glass** *(PMMA)*	Acrylglas *n*
6	**acrylonitrile butadiene styrene copolymers** *(ABS)*	Acrylnitril-Butadien-Styrol-Copolymere *npl*
7	**adapter**	Paßstück *n*
8	**additive**	Additiv *n*
9	**adhere, to**	haften
10	**adhesive**	Klebstoff *m*
11	**adhesive film**	Klebefolie *f*
12	**after-shrinkage**	Nachschwindung *f*
13	**ageing**	Alterung *f*
14	**ageing resistance**	Alterungsbeständigkeit *f*
15	**agglomerate**	Agglomerat *n*
16	**aliphatic**	aliphatisch

Français	*Español*	*Italiano*
ablatif	ablativo	ablativo
abrasion *f*	abrasión *f*	abrasione *f*
abrasif	abrasivo	abrasivo
accélérateurer *m*	acelerador *m*	acceleratore *m*
verre *m* acrylique	vidrio *m* acrílico	vetro *m* acrilico
copolymères *mpl* d'acrylonitrile-butadiène-styrène	copolímeros *mpl* de acrilonitrilo butadieno estireno	copolimeri *mpl* acrilonitrile butadiene stirene
adapteur *m*	adaptador *m*	adattore *m*
additif *m*	additivo *m*	additivo *m*
adhérer; coller	adherir; encolar; colar	aderire
adhésif *m*; colle *f*	adhesivo *m*; cola *f*	adesivo *m*
feuille *f* adhésive	película *f* adhesiva	pellicola *f* adesiva
postretrait *m*	contracción *f* posterior al moldeo	ritiro *m* dopo lo stampaggio
vieillissement *m*	envejecimiento *m*	invecchiamento *m*
tenue *f* au vieillissement	resistencia *f* al envejecimiento	resistenza *f* all'invecchiamento
aglomérat *m*	agglomeración *f*	agglomerato *m*
aliphatique	alifático	alifatico

Nr.	English	Deutsch
17	**alkyd resin**	Alkydharz *n*
18	**aluminium trihydrate**	Aluminiumtrihydrat *n*
19	**amorphous**	amorph
20	**anti-corrosive**	Korrosionsschutz
21	**antiblocking agent**	Antiblockmittel *n*
22	**antioxidant**	Antioxidans *n*
23	**antistatic agent**	Antistatikum *n*
24	**applicator roll**	Auftragwalze *f*
25	**aramid fibre**	Aramidfaser *f*
26	**atactic**	ataktisch
27	**austenitic**	austenitisch
28	**back pressure**	Staudruck *m*
29	**ball mill**	Kugelmühle *f*
30	**banbury mixer** *(internal mixer with plug)*	Banbury-Mischer *m*

Français	*Español*	*Italiano*
résine *f* alkyde	resina *f* alcídica	resina *f* alchidica
trihydrate *m* d'alumine	trihidrato *m* de alúmina	triidrati *mpl* di alluminio
amorphe	amorfo	amorfo
protection *f* anticorrosive	protección *f* frente a la corrosión	anticorrosivo *m*
agent *m* antiblocking	agente *m* antibloque	agente *m* antiaderenza; antibloccante *m*
antioxidant *m*	antioxidante *m*	antiossidante *m*
agent *m* antistatique	agente *m* antiestático	agente *m* antistatico
rouleau *m* enducteur	rodillo *m* aplicador	rullo *m* per applicazione
fibre *f* aramide	fibra *f* de aramida	fibra *f* aramidica
atactique	atáctico	attatico
austénitique	austenítico	austenitico
contre-pression *f*	contrapressión *f*	contropressione *f*
moulin *m* à billes	molino *m* de bolas	molino *m* a palle
mélangeur *m* banbury	mezclador *m* Banbury	mescolatore *m* banbury

Nr.	English	Deutsch
31	**barrel**	Zylinder *m*
32	**barrier plastic**	Barrierekunststoff *m*
33	**bead**	Schweißraupe *f*; Schweißwulst *m*
34	**bending**	Biegen *n*
35	**biaxial**	biaxial
36	**blade** *(of a mixer)*	Schaufel *f*
37	**blend**	Blend *n*; Abmischung *f*
38	**blister pack**	Blisterverpackung *f*
39	**block copolymer**	Blockcopolymer *n*
40	**blocking**	Blocken *n*
41	**blow moulding**	Blasformen *n*
42	**blow moulding line**	Blasformanlage *f*
43	**blowing mandrel**	Blasdorn *m*
44	**blowing mould**	Blasformwerkzeug *n*

Français	*Español*	*Italiano*
cylindre *m*	cilindro *m*	cilindro *m*; tamburo *m*
matière *f* plastique barrière	plástico *m* barrera	materia *f* plastica barriera
cordon *m* de soudure	cordón *m* de soldadura	cordone *m* di saldatura
cintrage *m*	curvado *m*	curvatura *f*
biaxiale	biaxial	biassiale
pale *f*	paleta *f*	pala *f*
blend *m*	mezcla *f*	mescola *f*
emballage *m* blistere	envasado *m* blister	imballagio *m* blister
copolymère *m* à blocs	copolímero *m* de bloque	copolimero *m* a bloque
blocking *m*	bloqueo *m*	bloccagio *m*
soufflage *m*	moldeo *m* por soplado	soffiatura *f*
souffleuse *f*	instalación *f* de soplado	line *f* di soffiaggio
mandrin *m* de soufflage	mandril *m* de soplado	mandrino *m* di soffiaggio
moule *m* de soufflage	molde *m* de soplado	stampo *m* di soffiaggio

Nr.	English	Deutsch
45	**blown film**	Blasfolie *f*
46	**breaker plate**	Lochscheibe *f*
47	**Brinell hardness**	Brinellhärte *f*
48	**bursting pressure**	Berstdruck *m*
49	**butadiene rubber**	Butadienkautschuk *m*
50	**butt welding**	Stumpfschweißen *n*
51	**cable sheathing**	Kabelummantelung *f*
52	**cadmium pigment**	Cadmiumpigment *n*
53	**calender**	Kalander *m*
54	**calendering**	Kalandrieren *n*
55	**calibrate, to**	kalibrieren
56	**calibrating device** *(pipes, profiles)*	Kalibriervorrichtung *f*
57	**carbon black**	Ruß *m*
58	**carbon fibre**	Kohlenstoffaser *f*
59	**cast film**	Gießfolie *f*

Français	Español	Italiano
film *m* soufflé	película *f* soplada	pellicola *f* soffiata
plaque *f* du filtre	placa *f* portadora	piastra *f* del filtro
dureté *f* Brinell	dureza *f* Brinell	durezza *f* Brinell
pression *f* d'éclatement	presión *f* de reventamiento	pressione *f* di scoppio
caoutchouc *m* butadiène	caucho *m* de butadieno	elastomero *m* butadiene
soudage *m* bout à bout	soldadura *f* a tope	saldatura *f* di testa
gainage *m* de câbles	cubierta *f* protectora de cable	protezione *f* per cavi
pigmento *m* di cadmio	pigment *m* de cadmium	pigmento *m* al cadmio
calandre *f*	calandria *f*	calandra *f*
calandrage *m*	calandrado *m*	calandratura *f*
calibrer	calibrar	calibrare
système *m* de calibrage	calibrador *m*	calibratore *m*
noir *m* de carbone	negro *m* de carbono	nerofumo *m*
fibre *f* de carbone	fibra *f* de carbona	fibra *f* di carbonio
film *m* coulé	película *f* colada	pellicola *f* colata

Nr.	English	Deutsch
60	**cast resin**	Gießharz *n*
61	**catalyst**	Katalysator *m*
62	**cavity**	Kavität *f*; Formnest *n*
63	**cavity pressure**	Werkzeuginnendruck *m*
64	**cellulose**	Cellusose *f*
65	**chelator**	Chelator *m*
66	**chemical resistance**	Chemikalienbeständig-keit *f*
67	**chill roll**	Kühlwalze *f*
68	**chopped glass fibre**	Kurzglasfasern *fpl*
69	**clamping force**	Formschließkraft *f*
70	**closed-cell** *(foam)*	geschlossenzellig
71	**co-rotating**	gleichläufig
72	**coating**	Beschichten *n*
73	**coating compounds**	Beschichtungsmassen *fpl*

Français	Español	Italiano
résine *f* de coulée	resina *f* de colada	resina *f* da colata
catalyseur *m*	catalizador *m*	catalizzatore *m*
cavité *f*; empreinte *f*	cavidad *f*	cavità *f*; impronta *f*
pression *f* interne du moule	presión *f* interior del molde	pressione *f* d'impronta
cellulose *f*	celulosa *f*	cellulosa *f*
agent *f* chelatant	quelante *m*	agente *m* chelante
résistance *f* chimique	resistencia *f* química	resistenza *f* chimica
cylindre *m* refroidisseau	cilindro *m* de enfriamiento	cilindro *m* di raffreddamento
fibres *fpl* courtes de verres	fibres *fpl* cortas de vidrio	fibras *fpl* cortas di vetro
force *f* de fermeture	presión *f* de cierre	forza *f* di chiusura
cellules *fpl* fermées	células *fpl* cerradas	celle *fpl* chiuse
corotatif	corrotativo	corotante
revêtement *m*; enduction *f*; couchage *m*	recubrimiento *m*	rivestimento *m*
produits *mpl* d'enduction	masas *fpl* de recubrimiento	mescole *fpl* per rivestimento

Nr.	English	Deutsch
74	**coefficient of expansion**	Ausdehnungskoeffizient *m*
75	**coextrusion**	Coextrusion *f*
76	**coextrusion line**	Coextrusionsanlage *f*
77	**cold forming**	Kaltumformen *n*
78	**cold slug**	kalter Pfropfen *m*
79	**collapsible core**	Faltkern *m*
80	**colourant**	Farbmittel *n*
81	**column**	Säule *f*; Holm *f*
82	**composite material**	Verbundwerkstoff *m*
83	**compound**	Compound *n*
84	**compounding**	Compoundieren *n*; Aufbereiten *n*
85	**compression moulding**	Preßformen *n*
86	**compression section** *(screw)*	Verdichtungszone *f*

Français	*Español*	*Italiano*
coefficient *f* de dilatation	coeficiente *m* de dilatación	coefficiente *m* di allungamento
coextrusion *f*	coextrusión *f*	coestrusione *f*
ligne *f* de coextrusion	instalación *f* de coextrusión	linea *f* di coestrusione
formage *m* à froid	moldeo *m* en frío	formatura *f* a freddo
goutte *f* froide; bouchon *m* froid	tapón *m* frio	punta *f* fredda
noyau *m* éclipsable	núcleo *m* colapsable	nucleo *m* a sprofondamento
colorant *m*	colorante *m*	colorante *m*
colonne *f*	columna *f*	colonna *f*
matériau *m* composite	material *m* laminado; composite *m*	materiale *m* compuesto
compound *m*	compound *m*	mescola *f*
compoundage *m*	preparación *f* de compuestos	mescolatura *f*
moulage *m* par compression	moldeo *m* por compresión	stampaggio *m* a compressione
zone *f* de compression	zona *f* de compresión	zona *f* di compressione

Nr.	English	Deutsch
87	**compressive strength**	Druckfestigkeit *f*
88	**compressive stress**	Druckspannung *f*
89	**computer aided design** *(CAD)*	rechnergestütztes Konstruieren *n*
90	**computer aided manufacturing** *(CAM)*	rechnergestützte Fertigung *f*
91	**computer integrated manufacturing** *(CIM)*	rechnerintegrierte Fertigung *f*
92	**continous filament**	Endlosfaser *f*
93	**conveyor**	Förderanlage *f*
94	**cooling jacket**	Kühlmantel *m*
95	**cooling mixer**	Kühlmischer *m*
96	**cooling rate**	Abkühlgeschwindigkeit *f*
97	**copolymer**	Copolymer *n*
98	**copolymersation**	Copolymerisation *f*

Français	*Español*	*Italiano*
résistance *f* à la compression *f*	resistencia *f* a la compresión	resistenza *f* alla compressione
contrainte *f* de compression	esfuersa *f* de compresión	sollecitazione *f* a compressione
conception *f* assistée par ordinateur *(CAO)*	construcción *f* asistida por ordenador	progettazione *f* con l'aiuto del calcolatore
fabrication *f* assistée par ordinateur *(FAO)*	fabricación *f* asistida por ordenator	produzione *f* controllata del calcolatore *(CAM)*
fabrication *f* intégrée par ordonateur *(CIM)*	fabricación *f* integrada	produzione *f* integrata con il calcolatore *(CIM)*
filament *m* continu	filamento *m* continuo	filamento *m* continuo
bande *f* transporteuse	transportador *m* mecánico	trasportatore *m*
double enveloppe *f* de refroidissement	camisa *f* de refrigeración	camicia *f* di raffreddamento
mélangeur *m* réfrigérant	mezclador *m* refrigerante	mescolatore *m* raffreddatore
vitesse *f* de refraichissement	velocidad *f* de enfriamiento	velocità *f* di raffreddatore
copoymère *m*	copolímero *m*	copolimero *m*
copolymérisation *f*	copolimerización *f*	copolimerizzazione *f*

Nr.	English	Deutsch
99	**corona treatment**	Coronabehandlung *f*
100	**counter-rotating**	gegenläufig
101	**coupling agent**	Haftvermittler *m*
102	**crack**	Riß *m*
103	**craze**	Craze *m*; Mikroriß *m*
104	**crazing**	Haarrißbildung *f*
105	**creep**	Kriechen *n*
106	**cure**	Aushärtung *f*
107	**cured**	ausgehärtet
108	**curing**	Härtung *f*
109	**curing time**	Härtezeit *f*
110	**cut**	Zuschnitt *m*
111	**cutting tool**	Schneidwerkzeug *n*
112	**cycle**	Zyklus *m*

Français	*Español*	*Italiano*
traitement *m* corona	tratamiento *m* corona	trattamento *m* corona
contra-rotative	giro contrario	controrotanti
agent *m* de couplage	agente *m* de adherencia	agente *m* di accoppiamento
fente *f*; fissure *f*; craquelure *f*	grieta *f*; fisura *f*	fessurazione *f*
craze *m*	microfisura *f*	screpolatura *f*
micro-feudillement *m*	formación *f* de grietas; cuarteamiento *m*	screpolatura *f*
fliage *m*	fluencia *f*	scorrimento *m*
cuisson *f*	endurecimiento *m*	indurimento *m*; trattamento *m*
réticulé	curado	reticolato
cuisson *f*	curación *f*	reticolazione *f*
temps *mpl* de cuisson	tiempo *m* de curado *m*	tempo *m* di cottura; tempo *m* di trattamento
flan *m*	corte *m*	ritaglio *m*
outil *m* de coupe	herramienta *f* de corte; cuchilla *f*	utensile *m* di pinzatura
cycle *m*	ciclo *m*	ciclo *m*

Nr.	English	Deutsch
113	**decompression section**	Dekompressionszone *f*
114	**deflashing**	Entgraten *n*
115	**deformation**	Deformation *f*
116	**degassing**	Entgasen *n*
117	**degradable**	abbaubar
118	**degradation**	Abbau *m*
119	**delamination**	Delaminierung *f*
120	**demould, to**	entformen
121	**depolymerisation**	Depolymerisation *f*
122	**die**	Düse *f*
123	**die gap**	Düsenspalt *m*
124	**die swell**	Düsenaufweitung *f*
125	**die-face pelletiser**	Heißabschlaggranulator *m*
126	**diecasting mould**	Druckgußwerkzeug *n*
127	**diffusion**	Diffusion *f*

Français	*Español*	*Italiano*
zone *f* de décompression	zona *f* de compresión	sezione *f* di decompressione
ébarbage *m*; ébavurage *m*	rebarbado *m*	sbavatura *f*
déformation *f*	deformación *f*	deformazione *f*
dégazage *m*	desgasificación *f*	degasaggio *m*
dégradable	degradable	degradabile
dégradation *f*	degradación *f*	degradazione *f*
délamination *f*	deslaminación *f*	delaminazione *f*
démouler	desmoldear	estrarre dallo stampo
dépolymérisation *f*	depolimerización *f*	depolimerizzazione *f*
filière *f*	boquilla *f*	filiera *f*
écartement *m* de filière	abertura *f* de la boquilla	apertura *f* de bocchettone
gouflement *m* de filière	dilatación *f* del *m* extrudado	rigonfiamento *m* nell'estrusione
granulatrice *f* á chaud	granuladora *f* con corte en caliente	granulatore *m* a caldo
moule de coulée *m*	molde *m* de colada	stampo *m* di colata
diffusion *f*	difusión *f*	diffusione *f*

Nr.	English	Deutsch
128	**dimension**	Abmessung *f*
129	**dip, to**	tauchen
130	**direct gating**	Direktanspritzen *n*
131	**disperse, to**	dispergieren
132	**dispersing agent**	Dispergierhilfsmittel *n*
133	**doctor knife**	Abstreifmesser *n*; Rakelmesser *n*
134	**doubling**	Kaschieren *n*
135	**dry blend**	Dryblend *n*
136	**dryer**	Trockner *m*
137	**ejection**	Auswerfen *n*
138	**ejector**	Auswerfer *m*
139	**ejector bolt**	Auswerferbolzen *m*
140	**ejector bush**	Auswerferhülse *f*
141	**ejector pin**	Auswerferstift *m*

Français	*Español*	*Italiano*
dimension *f*	dimensión *f*	dimensione *f*
tremper	sumergir	immergere
injection *f* directe	inyección *f* directa	orifizio *m* di entrata diretta
disperser	dispersar	disperdere
agent *m* dispersant	agente *m* dispersante	agente *m* di dispersione
racle *f*	cuchilla *f* fija; dosificadora *f* rasqueta	raschiatore *m*; racla *f*
doublage *m*	doublado *m*; recubrimiento *m* de una lámina con otra	accoppiamento *m*
mélange *m* à sec	mezcla *f* seca	mescolanza *f* secca
sécheur *m*	secadero *m*; secador *m*	essiccatore *m*
éjection *f*	expulsación *f*	espulsione *f*
éjecteur *m*	expulsor *m*	espulsore *m*
reçu *m* d'éjection	borno *m* extractor	bullone *m* dell'espulsore
éjection *f* tubulaire	manguito *m* del extractor	boccola *f* dell'espulsore
aigoulle *f* d'ejection	espiga *f* extractora	perno *m* espulsore

Nr.	English	Deutsch
142	ejector ring	Auswerferring *m*
143	ejector rod	Auswerferstange *f*
144	elastomer	Elastomer *n*
145	electroplating	Galvanisieren *n*
146	elongation at break	Bruchdehnung *f*
147	embossed sheeting	Prägefolie *f*
148	embossing	Prägen *n*
149	embossing calender	Prägekalander *m*
150	emulsifier	Emulgator *m*
151	emulsion	Emulsion *f*
152	engineering plastics	Konstruktionskunststoffe *fpl*
153	epoxy resin *(EP)*	Epoxidharz *n*
154	equipment *(machine)*	Ausrüstung *f*

Français	*Español*	*Italiano*
dévêtisseuse *f*	anillo *m* extractor	anello *m* dell'espulsore
tige *f* d'ejection	varilla *f* extractora	sbarra *f* dell'espulsore
élastomère *m*	elastómero *m*	elastomero *m*
galvanisation *f*	galvanización *f*	galvanoplastica *f*
allongement *m* à la ruptore	alargamiento *m* a la rotura	allungamente *m* a rottura
feuille *f* gaufrée	hoja *f* gofrada	foglia *f* goffrata
gaufrage *m*	estampado *m*; gofrado *m*	goffratura *f*
calandre *f* de gaufrage	calandra *f* de gofrado	calandra *f* di goffratura
emulsiomnant *m*	emulsionante *m*	emulsionante *m*
émulsion *f*	emulsión *f*	emulsione *f*
plastique *m* de construction; plastique *m* pour l'ingénierie	materiales *mpl* plásticos de construcción	tecnopolimeri *mpl*
résine *f* époxyde	resina *f* epoxi	resina *f* epossidica
équipement *m*; outillage *m*	equipo *m*; utillaje *m*	attrezzatura *f*; apparecchiatura *f*

Nr.	English	Deutsch
155	**ethylene vinyl acetate copolymer**	Ethylen-Vinylacetat-Copolymer *n*
156	**ethylene-propylene rubber**	Ethylen-Propylen-Kautschuk *m*
157	**expandable**	expandierbar
158	**extruder**	Extruder *m*
159	**extruder head**	Extruderkopf *m*
160	**extrusion**	Extrusion *f*
161	**extrusion blow moulding**	Extrusionsblasformen *n*
162	**extrusion die**	Extrusionswerkzeug *n*
163	**extrusion line**	Extrusionsanlage *f*
164	**extrusion welding**	Extrusionsschweißen *n*
165	**falling ball test**	Kugelfallversuch *m*
166	**falling dart test**	Fallbolzentest *m*
167	**feed**	Materialzufuhr *f*

Français	*Español*	*Italiano*
copolymère *m* d'ethylène-acétate de vinyle	copolímero *m* de etileno acetato de vinilo	copolimero *m* etileno acetato di vinile
caoutchouc *m* éthyléne-propyléne	caucho *m* de etileno-propileno	elastomero *m* etilene propilene
expansible	expandible	espandibile
extrudeuse *f*; boudineuse *f*	extrusora *f*	estrusore *m*
tête *f* d'extrusion	cabezal *m* de extrusión	testa *f* d'estrusione
extrusion *f*	extrusión *f*	extrusione *f*
extrusion-soufflage *f*	moldeo *m* por soplado	estrusione *f* soffiagie
filière *f*	boquilla *f* de extrusión	filiera *f* di estrusione
ligne *f* d'extrusion	línea *f* de extrusión	linea *f* di estrusione
soudage *m* par extrusion	soldeo *m* por extrusión	saldatura *f* per estrusione
essai *m* á la bille	ensayo *m* de caída de la bola	prova *f* caduta di sfera
test "dart" *m*	ensayo *m* daída del dardo	prova *f* di caduta con dardo
alimentation *f*	alimentación *f*	alimentazione *f*

Nr.	English	Deutsch
168	**feed hopper**	Einfülltrichter *m*; Aufgabetrichter *m*
169	**feed pump**	Dosierpumpe *f*
170	**feed ram**	Dosierkolben *m*
171	**feed roll**	Einzugswalze *f*
172	**feed section/zone**	Einzugszone *f*; Förderlänge *f*
173	**feed, to**	dosieren
174	**feeder; feeding equipment**	Dosiervorrichtung *f*
175	**fibre**	Faser *f*
176	**fibre composite**	Faserverbungwerkstoff *m*
177	**filler**	Füllstoff *m*
178	**film**	Folie *f*
179	**film blowing**	Folienblasen *n*
180	**film bubble**	Folienschlauch *m*

Français	Español	Italiano
trémie *f* d'alimentation	tolva *f* de alimentación	tramoggia *f* di carico
pompe *f* doseuse	bomba *f* de dosificación	pompa *f* d'alimentazione; pompa *f* dosatrice
doseure *m* á bande	dosificación *f* por émbolo	pistone *m* di alimentazione
cylindre *m* d'alimentation	cilindro *m* de alimentación	cilindro *m* di alimentazione
zone *f* d'alimentation	zona *f* de entrada; zona *f* de alimentación	zona *f* d'alimentazione
doser	dosificar	dosare; alimentare
dispositif *m* de dosage	dispositivo *m* de dosificación	dosatore *m*
fibre *f*	fibra *f*	fibra *f*
composite *m* á base de fibres	composite *m* a base de fibra	composito *m* con fibre
charge *f*	carga *f*	carica *f*
film *m*	película *f*; hoja *f*	pellicola *f*
soufflage *m* de gaine	soplado *m* de películas	filmatura *f* per soffiaggio
bulle *f* de film soufflé	burbuja *f* de film	bolla *f* di film

Nr.	English	Deutsch
181	**film gate**	Filmanguß *m*
182	**film streching plant**	Folienreckanlage *f*
183	**film web**	Folienbahn *f*
184	**film winder**	Folienwickler *m*
185	**finishing**	Nachbearbeitung *f*
186	**fit**	Passung *f*
187	**fixed platen**	Werkzeugaufspannplatte *f*, feststehende
188	**flame retardant**	Flammschutzmittel *n*
189	**flame treatment**	Beflammen *n*
190	**flame-spraying**	Flammspritzen *n*
191	**flammability**	Brennbarkeit *f*
192	**flash**	Grat *m*; Butzen *m*
193	**flat sheet film**	Flachfolie *f*
194	**flexographic printing**	Flexodruck *m*

Français	*Español*	*Italiano*
alimentation *f* en nappes	entrada *f* laminar	testa *f* di estrusione
ligne *f* de film orienté	instalación *f* de film orientado	impianto *m* di stiro di film
bande *f* de film	hoja *f* continua	nastro *m* di film
enrouleur *m*	bobinador *m*	avvolgitore *m*
finition *f*	acabado *m*; desbarbado *m*	finitura *f*
ajustage *m*	ajuste *m*	tolleranza *f*
plateaux *f* fixe du moule	placa *f* fija del molde	piastra *f* d'impronta fissa
ignifugeant *m* retardateur de combustion	producto *m* ignifugante	ritardante *m* la flamma
flammage *m*	tratamiento *m* con llama	trattamento *m* alla fiamma
flammage	proyección *f* a la llama; pulverización *f* a la llama	spruzzatura *f* a fiamma
inflammabilité *f*	inflamabilidad *f*	infiammabilità *f*
bavure *f*	rebaba *f*	bava *f*
feuille *f* à plat	làmina *f* plana	foglia *f* piana
flexographie *f*	flexografía *f*	flessografia *f*

Nr.	English	Deutsch
195	**flexural modulus**	Biegemodul *m*
196	**flexural stress**	Biegespannung *f*
197	**flight** *(of a screw)*	Gang *m*
198	**flocking**	Beflocken *n*
199	**fluid mixer**	Fluidmischer *m*
200	**fluidized bed coating**	Wirbelsintern *n*
201	**fluoropolymer**	Fluorpolymer *n*
202	**foam**	Schaumstoff *m*
203	**foam moulding**	Formschäumen *n*
204	**folding**	Abkanten *n*
205	**form-fill-seal machine**	Form-Füll-Siegelmaschine *f*
206	**fracture**	Bruch *m*

Français	*Español*	*Italiano*
module *m* de flexion	módulo *m* en flexión	modulo *m* a flessione
contrainte *f* en flexion	esfuerzo de flexión	sollecitazione *f* a flessione
pas *m*	hilo *m*; filete *m*	filetto *m*
flocage *m*	flocado *m*	floccaggio *m*
mélangeur *m* liquide	mezclador *m* de fluidos	mescolatore *m* liquido
sintérisation *f* en lit fluidere	recubrimiento *m* por inmersión en lecho fluidizado; sinterización *f* en lecho fluidizado	rivestimento *m* in letto fluidizzato
polymère *m* fluoré	polímero *m* flurocarbonado	polimero *m* fluorurato
mousse *f*	espuma *f*	schiuma *f*; espanso *m*
moulage *f* de mousse	moldeo *m* de espuma	stampaggio *m* di espanso
pliage *m*	doblado *m*	piegatura *f*
machine FFS *f*	máquina *f* de moldeo, llenado y sellado	macchina *f* formatrice riempitrice sigillatrice *(FFS)*
rupture *f*	fractura *f*	frattura *f*

Nr.	English	Deutsch
207	**friction welding**	Reibschweißen *n*
208	**gas injection moulding**	Gasinnendruckverfahren *n*
209	**gate**	Anschnitt *m*; Anguß *m*
210	**gel coat**	Gelcoatschicht *f*
211	**gel content**	Gelanteil *m*
212	**glass beads**	Glaskugeln *fpl*
213	**glass fibre**	Glasfaser *f*
214	**glass fibre mat**	Glasfasermatte *f*
215	**glass fibre reinforced plastics** *(GRP)*	Glasfaserkunststoff *m*; glasfaserverstärkter Kunststoff *m (GFK)*
216	**glass transition temperature**	Glastemperatur *f*
217	**graphite**	Graphit *m*
218	**grooved**	genutet
219	**grooved barrel extruder**	Nutenextruder *m*

Français	*Español*	*Italiano*
soudage *m* par friction	soldeo *m* por fricción	saldatura *f* per frizione
moulage *f* sous pression interne de gaz	moldeo *m* por inyección de gas	stampaggio *m* con prossione di gas
point *m* d'injection	entrada *f*	entrata *f*; orifizio *m* di entrata
gel coat *m*	gel coat *m*	finitura *f* superficiale
pourcentage *f* de produit gélifié	contenido *m* en gel	contenuto *m* di gel
billes *fpl* de verre	bolas *fpl* de vidrio	sferi *mpl* di vetro
fibre *f* de verre	fibra *f* de vidrio	fibra *f* di vetro
mat *m* de fibres de verre	mat *m* de vidrio	stuoia *f* di vetro; mat *m*
plastique *m* renforcé fibres de verre	plástico *m* reforzado con fibra de vidrio	materie *fpl* plastiche rinforzate con fibre di vetro
température *f* de transition vitreuse	temperatura *f* de transición vítrea	temperatura *f* di transizione vetrosa
graphite *m*	grafito *m*	grafite *f*
rainuré	ranurado	scanalto
extrudeuse *f* á picots	extrusora *f* con encamisado ranurado	estrusore *m* a tamburo scanalato

Nr.	English	Deutsch
220	**guide pin**	Führungsstift *m*; Führungssäule *f*
221	**hammer mill**	Hammermühle *f*
222	**hand lay-up**	Handlaminieren *n*
223	**hardener**	Härter *m*
224	**hardness test**	Härteprüfung *f*
225	**haul-off ratio**	Abzugsverhältnis *n*
226	**heat distortion temperature**	Formbeständigkeit *f* in der Wärme; Wärmeformbeständigkeit *f*
227	**heat resistant**	wärmeformbeständig
228	**heat-sealing**	Heißsiegeln *n*
229	**heated tool welding**	Heizelementschweißen *n*
230	**heating zone**	Heizzone *f*

Français	*Español*	*Italiano*
broche *f* de guidage	espiga *f* de guía; perno *m* de guía; columna *f* de guía	perno *m* di centratura; colonna *f* di guida
broyeure *m* á couteaux	molino *m* de mazos	mulino *m* a lame
drapage *f* manuel	moldeo *m* a mano	spalmatura *f* a mano
durcisseur *m*	endurecedor *m*	induritore *m*
essai *m* de dureté	ensayo *m* de dureza	prova *f* di durezza
taux *m* de tirage	relación *f* de estiraje	percentuale *f* di traino
température *f* de déformation à la chaleur	temperatura *f* de distorsión por el calor	temperatura *f* di distorsione sotto carico e calore
résistant à la chaleur; thermostable	resistente al calor	termoresistente
soudage *m* à chaud	soldeo *m* por calefacción; sellado *m* en caliente	termosaldatura *f*
soudage *m* par éléments chauffants	soldadeo por elementos de calefacción	saldatura *f* con utensile caldo
zone *f* de chauffage	zona *f* de calefacción	zona *f* di riscaldamento

Nr.	English	Deutsch
231	**high-frequency welding**	Hochfrequenzschweißen *n*
232	**holding pressure**	Nachdruck *m*
233	**homopolymer**	Homopolymer *n*
234	**hopper**	Trichter *m*
235	**hot runner**	Heißkanal *m*
236	**hot runner mould**	Heißkanalwerkzeug *n*
237	**hot runner nozzle**	Heißkanaldüse *f*
238	**hot-gas welding**	Heißgasschweißen *n*; Warmgasschweißen *n*
239	**hydrolysis**	Hydrolyse *f*
240	**impact strength**	Schlagzähigkeit *f*
241	**impregnate, to**	imprägnieren
242	**impulse welding**	Impulsschweißen *n*
243	**impulse welding** *(thermal)*	Wärmeimpulsschweißen *n*

Français	Español	Italiano
soudage *m* à haute fréquence	soldeo *m* por alta frecuencia	saldatura *f* ad alta frequenza
pression *f* de maintien	presión *f* posterior	pressione *f* di mantenimento
homopolymère *m*	homopolímero *m*	omopolimero *m*
trémie *f*	tolva *f* mentación; tolva *f*	tramoggia *f*
canal *m* chauffant	canal *m* caliente	canale *m* caldo
moule *m* à canal chauffant	molde *m* con canal caliente	stampo *m* a canali coldi
buse *f* pour canal chaud	boquilla *f* del canal caliente	ugello *m* a canale caldo
soudage *m* aux gaz chauds	soldeo *m* por gas caliente	saldatura *f* ad aria calda
hydrolyse *f*	hidrólisis *f*	idrolisi *f*
résistance *f* au choc	resistencia *f* al impacto	resistenza *f* all'urto
imprégner	impregnar	impregnare
soudage *f* pour impulsion	soldeo *m* por impulsión	saldatura *f* a impulsi
soudage *m* par impulsions thermiques	soldeo *m* por impulso térmico	termosaldatura *f* ad impulsi

Nr.	English	Deutsch
244	**in-situ foam**	Ortsschaum *m*
245	**inhibitor**	Inhibitor *m*
246	**injection blow moulding**	Spritzblasen *n*
247	**injection compression moulding**	Spritzprägen *n*
248	**injection mould**	Spritzgießwerkzeug *n*
249	**injection moulded part**	Spritzgußteil *n*
250	**injection moulding**	Spritzgießen *n*
251	**injection moulding machine**	Spritzgießmaschine *f*
252	**injection nozzle**	Einspritzdüse *f*
253	**injection pressure**	Einspritzdruck *m*
254	**injection unit**	Spritzeinheit *f*
255	**insert**	Einpreßteil *n*; Einlegeteil *n*

Français	*Español*	*Italiano*
mousse *f* in situ	espumación *f* "in situ"	espanso *m* in loco
inhibiteure *m*	inhibidor *m*	inibitore *m*
injection *f* soufflage	inyección *f* soplado	stampaggio *m* per soffiagio a iniezione
moulage *m* par injection-compression	inyección *f* compréssion	stampaggio *m* per compressione a iniezione
moule *m* à injection *f*	molde *m* para inyección	stampo *m* per iniezione
pièce *f* moulée par injection	pieza *f* inyectada	pezzo *m* stampato per iniezione
moulage *m* par injection	moldeo *m* por inyección	stampaggio *m* ad iniezione
machine *f* à injecter	máquina *f* de moldeo por inyección; inyectora *f*	pressa *f* per iniezione
buse *f* d'injection	boquilla *f* de inyección	ugello *m* di iniezione
pression *f* d'injection	presión *f* de inyección	pressione *f* d'iniezione
unité *f* d'injection	unidad *f* de inyección	unità *f* di iniezione
insert *f*	inserto *m*	inserto *m*

Nr.	English	Deutsch
256	**insulating varnish**	Isolierlack *m*
257	**insulting material**	Dämmstoff *m*
258	**interlaminar**	interlaminar
259	**internal cooling**	Innenkühlung *f*
260	**internal mixer**	Innenmischer *m*
261	**internal pressure**	Innendruck *m*
262	**ionomer**	Ionomer *n*
263	**irradiation**	Bestrahlung *f*
264	**isotactic**	isotaktisch
265	**jacket**	Mantel *m*
266	**kneader**	Kneter *m*
267	**laminate**	Schichtstoff *m*; Laminat *n*
268	**lap-welding**	Überlappschweißen *n*
269	**lead stabiliser**	Bleistabilisator *m*
270	**leakage flow**	Leckströmung *f*

Français	*Español*	*Italiano*
vernis *m* d'isolation	barniz *m* aislante	vernice *f* isolante
matériaux *m* isolant	material *m* disiante	materiale *m* isolante
interlaminaire	interlaminar	interlaminare
refraichissement *m* interne	enfriamiento *m* interno	raffreddamento *m* interno
mélangeur *m* interne	mezclador *m* interno	mescolatore *m* interno
pression *f* interne	presión *f* interna	pressione *f* interna
ionomère *m*	ionómero *m*	ionomero *m*
irridation *f*	irradiación *f*	irradiazione *f*
isotactique	isotáctico	isotattico
enveloppe *f*	camisa *f*	camicia *f*
malaxeur *m*	amasadora *f*; amasador *m*	impastatrice *f*
stratifié *m*; lamifié *m*	laminado *m*	stratificato *m*
soudage *m* par recouvrement	soldeo *f* a solape	saldatura *f* a sovrapposizione
stabilisant *m* au plomb	estabilizador *m* de plomo	stabilizzante *m* al piombo
fuite *f*	flujo *m* de pérdidas	flusso *m* di ruggito

Nr.	English	Deutsch
271	**leakage test**	Dichtheitsprüfung *f*
272	**light stabiliser**	Lichtschutzmittel *n*
273	**lining**	Auskleidung *f*
274	**locking mechanism**	Arretiervorrichtung *f*
275	**low pressure process**	Niederdruckverfahren *n*
276	**lubricant**	Gleitmittel *n*
277	**machine, to**	bearbeiten
278	**macromolecule**	Makromolekül *n*
279	**mandrel**	Pinole *f*
280	**mandrel support**	Dornhalter *m*
281	**manifold**	Verteiler *m*; Schmelzeverteiler *m*
282	**masterbatch**	Masterbatch *n*
283	**mat reinforcement**	Mattenverstärkung *f*

Français	Español	Italiano
test *f* d'étanchéité	ensayo *m* de estanqueidad	prova *f* di tenuta
stabilisant *m* á la lumière	estabilización *f* a la luz	stabilizzante *m* alla luce
chemise *f*	revestimiento *m*	rivestimento *m*
mécanisme *m* de formeture	mecansismo *m* de cierre	meccanismo *m* di chiusura
procédé *m* basse pression	proceso *m* de baja presión	processo *m* a bassa pressione
lubrifiant *m*	lubricante *m*	lubrificante *m*
usiner	mecanizar; maquinar	lavorare all'utensile
macromolécule *f*	macromolécula *f*	macromolecola *f*
mandrin *m*; douille *f*	mandril *m*	mandrino *m*
porte-poinçon *m*	portamandril *m*	anello *m* a razze
repartiteur	distribuidor *m*	collettore *m*
mélange-maître *m*	mezcla *f* básica; mezcla *f* madre	mescolanza *f* madre
mat *m* de renfort	mat *m* de refuerzo	mat *m* di rinforzo

Nr.	English	Deutsch
284	**maximum allowable concentration** *(MAC)*	Maximale Arbeitsplatzkonzentration *f (MAK-Wert m)*
285	**melamine resin**	Melaminharz *n*
286	**melt**	Schmelze *f*
287	**melt flow index** *(MFI)*	Schmelzindex *m*
288	**melting point**	Schmelzpunkt *m*
289	**melting temperature**	Schmelztemperatur *f*
290	**meter, to**	dosieren
291	**metering pump**	Dosierpumpe *f*
292	**mill**	Mühle *f*
293	**mineral filler**	Mineralfüllstoff *m*
294	**mixer**	Mischer *m*
295	**mixing head**	Mischkopf *m*
296	**modifier**	Modifiziermittel *n*
297	**modulus**	Modul *m*

Français	Español	Italiano
concentration *f* maximum sur le lieu de travail	concentración *f* máxima admisible *(MCA)*	concentrazione *f* massima consentita *(CMC)*
mélamine *f*	resina *f* de melamina	resina *f* melaminica
matière *f* fardue	masa *f* fundida	materiale *f* fuso
indice *m* de fusion *f*	indice *m* de fusión	indice di fluidità a caldo
point *m* de fusion	punto *m* de fusión	punto *m* di fusione
température *f* de fusion	temperatura *f* de fusión	temperatura *f* di fusione
doser	dosificar	dosare; alimentare
pompe *f* doseuse	bomba *f* de dosificación	pompa *f* d'alimentazione; pompa *f* dosatrice
broyeur *m*	molino *m*	frantumatore *m*
charge *f* minérale	carga *f* mineral	carica *f* minerale
mélangeur *m*	mezclador *m*	mescolatrice *f*
tête *f* de mélange	cabezal *m* mezclador	testa *f* di miscelazione
modifiant *m*	modificador *m*	modificante *m*
module *m*	módulo *m*	modulo *m*

Nr.	English	Deutsch
298	**modulus of elasticity**	Elastitzitätsmodul *m*
299	**moisture content**	Feuchtigkeitsgehalt *m*
300	**monofilament**	Monofil *n*
301	**monomer**	Monomer *n*
302	**morphology**	Morphologie *f*
303	**mould**	Spritzgießwerkzeug *n*
304	**mould release agent**	Formtrennmittel *n*
305	**mould, to**	formen; spritzgießen
306	**moulding**	Formteil *n*
307	**moulding compounds**	Formmasse *f*
308	**moulding plant**	Spritzgießbetrieb *m*
309	**moving mould half**	Schließseite *f*
310	**moving platen**	Werkzeugaufspannplatte *f*, bewegliche

Français	Español	Italiano
module *m* d'eléctricité	módulo *m* de elasticidad	modulo *m* di elasticità
cause *f* d'humidité	contenido *m* en humedad	contenuto *m* d'umidità
monofilament *m*	monofilamento *m*	monofilamento *m*
monomére *m*	monómero *m*	monomero *m*
morphologie *f*	morfología *f*	morfologia *f*
moule *m*	molde *m* de inyección	stampo *m*
agent *m* de démoulage	agente *m* de desmoldeo	agente *m* antiadesivo; distaccante *m*
mouler	moldear	stampare
pièce *f* moulèe	pieza *f* moldeada	pezzo *m* stampato
compound *m* à mouler	material *f* de moldeo	materiali *m* plastici per stampaggio
atelier *m* d'injection	taller *m* de moldeo	fabbrica *f* di stampaggio
côté *m* du plateau mobile	lado *m* del semimolde móvil	piano *m* mobile
plateau *f* mobile du moule	placa *f* móvil del molde	piastra *f* mobile

Nr.	English	Deutsch
311	**multi-cavity mould**	Mehrfachwerkzeug *n*
312	**natural rubber**	Naturkautschuk *m*
313	**needle valve**	Nadelventil *n*; Nadelverschluß *m*
314	**nip**	Walzenspalt *m*
315	**nip rolls**	Abquetschwalzen *fpl*
316	**non-woven**	Vlies *n*; Vliesstoff *m*
317	**notched**	gekerbt
318	**nozzle**	Düse *f*
319	**nucleating agents**	Nukleierungsmittel *n*; Keimbildner *m*
320	**oligomer**	Oligomer *n*
321	**output rate**	Durchsatzleistung *f*
322	**outsert moulding**	Outsert-Technik *f*
323	**parison**	Vorformling *m*

Français	Español	Italiano
mulitmoule *m* à empreintes	molde *m* de varias posiciones; molde *m* múltiple	stampo *m* multiplo; stampo *m* a più cavità
caoutchouc *m* naturel	caucho *m* natural	gomma *f* naturale
clapet *m* à aiguille	válvula *f* de aguja	valvola *f* a spillo
espace *m* entre cylindres	abertura *f* entre cilindros	scarto *m* fra rulli
rouleaux *mpl* pinceurs	rodillos *mpl* de compresión	rulli *mpl* di traino
nontissé *m*	velo *m*	tessuto *m* non tessuto
entaillé	entalla	con intaglio
buse *f*	boquilla *f*	ugello *m*
agent *m* nucléant	agentes *mpl* nucleantes	germinatori *mpl*; nucleizzatori *mpl*
oligomère *m*	oligomero *m*	oligomero *m*
niveau *m* de débit	rendimiento *m* de producción	livello *m* di produzione
outsert moulding *m*	técnica "outsert"	stampaggio *m* "outsert"
paraison *f*; ébauché *f*	preforma *f*; parisón *m*	parison *m*; preforma *f*

Nr.	English	Deutsch
324	**particle**	Partikel *f*
325	**paste**	Paste *f*
326	**pellet**	Granulat *n*
327	**pelletise, to**	granulieren
328	**pendulum impact test**	Pendelschlagversuch *m*
329	**perforated plate**	Lochplatte *f*
330	**phenol formalde resin** *(PF)*	Phenol-Formaldehyd-Harz *n*
331	**pigment**	Pigment *n*
332	**pilot plant**	Pilotanlage *f*
333	**pin gate**	Punktanguß *m*
334	**plasticator**	Plastifiziereinheit *f*
335	**plasticising capacity**	Plastifizierleistung *f*
336	**plasticising section** *(of a screw)*	Plastifizierzone *f*
337	**plasticizer**	Weichmacher *m*
338	**plastics**	Kunststoffe *mpl*

Français	Español	Italiano
particule *f*	partícula *f*	particella *f*
pâte *f*	pasta *f*	pasta *f*
granulé *m*	granulado *m*	granulo *m*
granuler	granular	granulare
test *m* au choc du pendule	ensayo *m* de impacio por péndola	prova *f* d'urto con pendolo
plaque *f* perforée	placa *f* perforada	piastra *f* perforata
résine *f* phénol formaldéhyde	resina *f* de fenol-formol	resina *f* fenol formaldeide
pigment *m*	pigmento *m*	pigmento *m*
unité *f* pilote	instalación *f* piloto	impianto *m* pilota
injection *f* capillaire	bebedero *m* capilar	orifizio *m* di entrata a punta di spillo
plastificateur *m*	plastificador *m*	plastificante *m*
capacité *f* de plastification	capacidad *f* de plastificación	capacità *f* di plastificazione
zone *f* de plastification	zona *f* de plastificación	zone *f* di plastificazione
plastifiant *m*	plastificante *m*	plastificante *m*
matière *f* plastique; plastique *m*	plásticos *mpl*	materie *fpl* plastiche

Nr.	English	Deutsch
339	**plastisol**	Plastisol *n*
340	**platen**	Werkzeugträger *m*; Aufspannplatte *f*
341	**plunger**	Kolben *m*
342	**polishing**	Polieren *n*
343	**polishing roll**	Glättwalze *f*
344	**polyacetal** *(POM)*	Polyacetal *n*
345	**polyamide** *(PA)*	Polyamid *n*
346	**polyamideimide** *(PAI)*	Polyamidimid *n*
347	**polybutylene terephthalate** *(PBT)*	Polybutylenterephthalat *n*
348	**polycarbonate** *(PC)*	Polycarbonat *n*
349	**polyester**	Polyester *n*
350	**polyether**	Polyether *n*
351	**polyether etherketone** *(PEEK)*	Polyetheretherketon *n*
352	**polyether imide** *(PEI)*	Polyetherimid *n*
353	**polyethersulphone** *(PES, PESU)*	Polyethersulfon *n*
354	**polyethylene** *(PE)*	Polyethylen *n*

Français	*Español*	*Italiano*
plastisol *m*	plastisol *m*	plastisol *m*
plateau *m* porte-moule	placa *f* portamolde	piastra *f*; piano *m* di pressa
plongeur *m*	pistón *m*	pistone *m*
polissage *m*	pulimento *m*	lucidatura *f*
cylindre *m* de polissage	rodillo *m* de satinado	rullo *m* di finitura
polyacétal *m*	poliacetal *m*	poliacetale *m*
polyamide *m*	poliamida *f*	poliamide *m*
polyamide-imide *m*	poliamidimida *f*	poliamideimide *m*
polybutylène-téréphtalate *m*	politereftalato *m* de butilo	polibutilenterefta-lato *m*
polycarbonate *m*	policarbonato *m*	policarbonato *m*
polyester *m*	poliéster *m*	poliestere *m*
polyéther *m*	poliéter *m*	polietere *m*
Polyétheréther-cétone *m*	poliéteretercetona *f*	polietere-etere-chetone *m*
polyétherimide *m*	poliéterimida *f*	polieterimmide *f*
polyéthersulfone *m*	poliétersulfona *f*	polietersulfone *m*
Polyéthylène *m*	polietileno *m*	polietilene *m*

Nr.	English	Deutsch
355	**polyethylene terephthalate** *(PET)*	Polyethylenterephthalat *n*
356	**polymethylmethacrylate** *(PMMA)*	Polymethylmethacrylat *n*
357	**polyphenylene ether** *(PPE)*, **modified**	Polyphenylenether, modifiziert *n*
358	**polypropylene** *(PP)*	Polypropylen *n*
359	**polystyrene** *(PS)*	Polystyrol *n*
360	**polysulfone** *(PSU)*	Polysulfon *n*
361	**polytetrafluoro ethylene** *(PTFE)*	Polytetrafluorethylen *n*
362	**polyurethane** *(PU)*	Polyurethan *n*
363	**polyvinyl chloride** *(PVC)*	Polyvinylchlorid *n*
364	**post-curing**	Nachhärten *n*
365	**post-shrinkage**	Nachschwindung *f*
366	**pre-heating**	Vorwärmen *n*
367	**prepreg**	Prepreg *m*
368	**press**	Presse *f*
369	**printing**	Bedrucken *n*

Français	*Español*	*Italiano*
polyéthylènetér-éphtalate *m*	politereftalato *m* de etileno	polietilentereftalato *m*
polyméthylmétha-crylate *m*	polimetacrilato de polimetilo *m*	polimetilmeta-crilato *m*
polyéther *m* de phenylène, modifié	poliéter de fenileno, modificado	polifeniletere *m*, modificato
polypropylène *m*	polipropileno *m*	polipropilene *m*
polystyrène *m*	poliestireno *m*	polistirene *m*
polysulfone *m*	polisulfona *f*	polisolfone *m*
polytétrafluor-éthylène *m*	politetrafluoretileno *m*	politetrafluoro-etilene *m*
polyuréthanne *m*	poliuretano *m*	poliuretano *m*
chlorure *m* de polyvinyle	policloruro de vinilo *m*	cloruro *m* di polivinile
post cuisson *f*	post curado *m*	post-trattamento *m*
postretrait *m*	contracción *f* posterior al moldeo	ritiro *m* dopo lo stampaggio
préchauffage *m*	precalentamiento *m*	preriscaldamento *m*
préimprégné *m*	preimpregnado *m*	preimpregnato *m*
presse *f*	prensa *f*	pressa *f*
impression *f*	imprimir *f*	stampa *f*

Nr.	English	Deutsch
370	**processor**	Verarbeiter *m*
371	**property**	Eigenschaft *f*
372	**pultrusion**	Pultrusion *f*
373	**ram extruder**	Kolbenextruder *m*; Ramextruder *m*
374	**rate of heating**	Aufheizrate *f*
375	**raw material**	Rohstoff *m*
376	**reaction injection moulding** *(RIM)*	Reaktionsspritzgießen *n*
377	**recycle, to**	wiederaufarbeiten
378	**reinforced plastics**	verstärkte Kunststoffe *mpl*
379	**rejects**	Ausschuß *m*
380	**resin**	Harz *n*
381	**resin impregnated**	harzimprägniert
382	**rigid**	hart
383	**rigid foam**	Hartschaumstoff *m*

Français	*Español*	*Italiano*
transformateur *m*	transformador *m*	trasformatore *m*
propriété *f*	propiedad *f*	proprietà *f*
pultrusion *f*	pultrusión *f*	pultrusione *f*; estrusione *f* per tirata
extrudeuse *f* à piston	prensa *f* de extrusión con émbolo	estrusore *m* a pistone
taux *m* de chauffage	factor *m* de calefacción	tasso *m* di riscaldamento
matière *f* première	material *m* prima	materia *f* prima
moulage *m* par injection-réaction	moldeo *m* por inyección reactiva	stampaggio *m* per reazione iniezione
recycler	recuperar	rilavorare; riciclare
plastiques *mpl* renforcés	plásticos *mpl* reforzados	materie *fpl* plastiche rinforzate
rebut *m*	desperdicio *m*	scarti *mpl*
résine *f*	resina *f*	resina *f*
imprégnée de résine	impregnato de resina	impregnato con resina
rigide	rígido	rigido
mousse *f* rigide	espuma *f* rígida	espanso *m* rigido

Nr.	English	Deutsch
384	**roller coating**	Walzenauftrag *m*
385	**rotary table**	Drehtisch *m*
386	**rotate, to**	rotieren
387	**rotational moulding**	Rotationsformen *n*
388	**roving**	Glasfaserstrang *m*
389	**rubber**	Kautschuk *m*; Gummi *m*
390	**runner**	Angußkanal *m*
391	**screen**	Filtersieb *n*
392	**screw**	Schnecke *f*
393	**sealing**	Siegeln *n*
394	**semi-finished products**	Halbzeug *n*
395	**semi-rigid**	halbhart
396	**sheeting, sheet**	Platte *f*
397	**shot volume**	Spritzvolumen *n*
398	**shrinkage**	Schwindung *f*

Français	*Español*	*Italiano*
enduction *f* à rouleaux	recubrimiento *m* con rodillo	rivestimento *m* a rullo
table *f* rotative	mesa *f* giratoria	tavola *f* rotante
tourner	girar	ruotare
moulage *m* par rotation	moldeo *m* rotacional	formatura *f* rotazionale
roving *m*	roving *m*	roving *m*; stoppino *m*
caoutchouc *m*	caucho *m*; goma *f*	gomma *f*
canal *m* d'alimentation	canal *m* de alimentación	canale *m* principale
filtre *m*	filtro *m*	filtro *m*
vis *f*	tornillo *m*	vite *f*
soudure *f*	sellado *m*	saldatura *f*
demi-produit *m*	producto *m* semiacabado; semiacabado *m*	semilavorato *m*
semirigide	semirígido	smirigido
feuille *f*	hoja *f*; lamina *f*	lastra *f*; foglia *f*
volume *m* injectable	volumen *m* inyectado	volume *m* d'iniezione
retrait *m*	contracción *f*	ritiro *m*

Nr.	English	Deutsch
399	**silk screen printing**	Siebdruck *m*
400	**single screw**	Einfachschnecke *f*
401	**single-cavity mould**	Einfachwerkzeug *n*
402	**single-flighted** *(screw)*	eingängig
403	**single-screw extruder**	Einschneckenextruder *m*
404	**sliding friction**	Gleitreibung *f*
405	**slot die**	Breitschlitzdüse *f*
406	**slush mould** *(plastisols)*	Gießwerkzeug *f*
407	**softening point**	Erweichungspunkt *m*
408	**solvent**	Lösungsmittel *n*
409	**split mould**	Backenwerkzeug *n*
410	**spot welding**	Punktschweißen *n*
411	**sprue**	Anguß *m*

Français	Español	Italiano
sérigraphie	serigrafía *f*	serigrafia *f*
mono vis *f*	tornillo *m* único	monovite *f*
moule *m* mono empreinte	molde *m* de una sola cavidad	stampo *m* a una impronta
filet m, á un seul	tornillo *m* de un sólo filete	vite *f* con un solo filetto
extrudeuse *f* mono vis	extrusora *m* de un tornillo	estrusore *m* monovite
frottement *m* dynamique	fricciòn *f* destizante	frizione *f* di scorrimento
filière *f* plate	boquilla *f* de plana	filiera *f* piana; testa *f* piana
moule *m* à plastisols	molde *m* de embarrado *(plastisoles)*	forma *f* per plastisol
température *f* de ramollissement	temperatura *f* de reblandecimiento	temperatura *f* di rammollimento
solvant *m*	disolvente *m*	solvente *m*
moule *m* à coins	molde *m* partido	stampo *m* diviso
soudure *f* par points	soldeo *f* por puntos	saldatura *f* a punti
carotte *f*	bebedero *m*	materozza *f*; canale *m* principale di iniezione

Nr.	English	Deutsch
412	**sprue bush**	Angußbuchse *f*
413	**sprueless**	angußfrei
414	**stabilizer**	Stabilisator *m*
415	**stack mould**	Etagenwerkzeug *n*
416	**standard test piece**	Normstab *m*
417	**standardization**	Typisierung *f*
418	**start, to**	anfahren
419	**static friction**	Haftreibung *f*
420	**stationary**	feststehend
421	**stop pin**	Anschlagbolzen
422	**strain**	Dehnung *f*
423	**strainer**	Lochscheibe *f*
424	**strength**	Festigkeit *f*
425	**stress**	Spannung *f*
426	**stretch, to**	verstrecken

Français	*Español*	*Italiano*
douille *f* de carotte	manguito *m* del bebedero	boccola *f* di iniezione
sans carotte *f*	sin bebedero *m*	senza carota *f*
stabilisant *m*	estabilizador *m*	stabilizzante *m*
moule *m* à étages	molde *m* de pisos	stampo *m* a piani
prouvette *f* normalisée	probeta *f* normalizada	campione *m* normalizzato
normalisation *f*; standardisation *f*	normalización *f*; estandardización *f*	unificazione *f*
faire démarrer	poner en marcha	avviare; mettere in funzione
frottement *m* statique	rozamiento *m* de adherencia	frizione *f* statica
fixe	fijo estacionario	fisso
obturateur *m* à aiguille	perno *m* de tope	perno *m* di arresto
dilatation *f*	alargamiento *m*	deformazione *f*
grille *f*	plato *m* rompedor	piastra *f* perforata
résistance *f*	resistencia *f*	resistenza *f*
tension *f*	fuerza *f*	sollecitazione *f*
étirer	estirar	stirare

Nr.	English	Deutsch
427	**stripper plate**	Abstreifplatte *f*
428	**stroke**	Hub *m*
429	**structural foam**	Integralschaumstoff *m*
430	**styrene acrylonitrile copolymers** *(SAN)*	Styrol-Acrylinil-Copolymer *n*
431	**styrene butadiene copolymer** *(SB)*	Styrol-Butadien-Copolymer *n*
432	**swivel-out**	ausschwenkbar
433	**take-off rolls**	Abzugswalze *f*
434	**take-off speed**	Abzugsgeschwindigkeit *f*
435	**take-off unit**	Abzugsvorrichtung *f*
436	**talc**	Talkum *n*
437	**tempering**	Tempern *n*
438	**test**	Prüfung *f*
439	**test specimen**	Probekörper *m*
440	**thermoforming**	Warmformen *n*

Français	*Español*	*Italiano*
plaque *f* d'extraction	placa *f* extractora	piastra *f* di smontaggio
course *f*	recorrido *m*; carrera *f*	corsa *f*
mousse *f* structurale	espuma *f* integral	espanso *m* strutturale
copolymère *m* de styrène-acrylonitrile	copolímero *m* de estireno-acrilonitrilo	copolimero *m* stirene acrilonitrile
copolymère *m* de styrène butadiène	copolímero *m* de estireno butadieno	copolimero *m* stirene butadiene
pivotant	basculante	tavola rotante
cylindre *m* de tirage; derouleur *m*	rodillo *m* de tren	rulli *mpl* di traino
vitesse *f* de tirage *m*	velocidad *f* de estiraje	velocità *f* di stiro
unité *f* de tirage	tren *m* de arrastre	unità *f* di tirata
talc *m*	talco *m*	talco *m*
recuit *m*	recocido *m*	rinvenimento *m*
essai *m*	ensayo *m*	prova *f*
éprouvette *f*	probeta *f*	provetta *f*
thermoformage *m*	termoconformado *m*	termoformatura *f*

Nr.	English	Deutsch
441	**thermoplastic elastomers**	thermoplastische Elastomere *npl*
442	**thermoplastic resin**	Thermoplast *m*
443	**thermoset**	Duroplast *m*
444	**thermosetting**	hitzehärtbar
445	**thickness gauge**	Dickenmeßeinrichtung *f*
446	**thread**	Gewinde *n*
447	**three-plate mould**	Dreiplattenwerkzeug *n*
448	**three-roll calender**	Dreiwalzenkalander *m*
449	**toggle**	Kniehebel *m*
450	**torpedo**	Torpedo *m*
451	**transfer mould**	Spritzpreßwerkzeug *n*
452	**transfer moulding**	Spritzpressen *n*
453	**transfer moulding press**	Spritzpresse *f*

Français	Español	Italiano
élastomères *mpl* thermoplastiques	elastómeros *mpl* termoplásticos	elastomeri *mpl* termoplastici
thermoplastique *m*	termoplástico *m*	materiale *m* termoplastico
thermodurcissable *m*	plástico *m* termoestable	termoindurente *m*
thermodurcissable	termoendurecible	termoindurente
gauge *f* d'épaisseur	medidor *m* de espesor	spessore *m*
fil *m*	filete *m*	fil *m*
moule *m* bi-étage *m*	molde *m* de tres placas	stampo *m* con piastra intermedia
calandre *f* à trois cylindres	calandra *f* de tres cilíndros	calandra *f* a tre cilindri
genouillère *f*	palanca *f* acodada	ginocchiera *f*
torpille *f*	torpedo *m*	testa *f*; siluro *m*
moule *m* à transfert	molde *m* de transferencia	stampo *m* trasferimento
moulage *m* par transfer	moldeo *m* por transferencia	pressa *f* per stampaggio a trasferimento
presse *f* transfer	prensa *f* de moldeo por transferencia	stampaggio *m* mediante trasferimento

Nr.	English	Deutsch
454	**triple screw**	Dreifachschnecke *f*
455	**tubular die**	Ringdüse *f*
456	**twin-screw extruder**	Doppelschneckenextruder *m*
457	**undercut**	Hinterschneidung *f*
458	**uniaxial**	einachsig
459	**UV-stabilizer**	UV-Stabilisator *m*
460	**vacuum forming**	Vakuumverformung *f*; Tiefziehen *n*
461	**vent zone**	Entgasungszone *f*
462	**venting channel**	Entlüftungskanal *m*
463	**viscosity**	Viskosität *f*
464	**void**	Lunker *m*
465	**warpage**	Verzug *m*
466	**waste**	Abfall *m*
467	**water treeing**	Bäumchenbildung *f*

Français	*Español*	*Italiano*
triple vis *f*	triple tornillo *m*	vite *f* tripla
filière *f* annulaire	tobera *f* anular	filiera *f* anulare
extrudeuse *f* double vis	extrusora *m* de dos tornillos	estrusore *m* a doppia vite
contre-dépouille *f*	contrasalida *f*; rebaja *f*	sottosquadro *m*
uniaxial	uniaxial	monoassiale
stabilisant *m* UV	estabilizador *m* UV	stabillizzante *m* UV
formage *m* sous vide *m*	moldeo *m* por vacío	formatura *f* sotto vuoto
évent *m*	zona *f* de desgasificación	zona *f* di ventilazione
trou *m* d'évent	salida *m* de aire; respiradero *m* per l'aria	sfogo *m* per l'aria
viscosité *f*	viscosidad *f*	viscosità *f*
bulle *f*	poro *m*	soffiatura *f*; risucchio *m*
gauchissement *m*	deformación *f*	deformazione *f*
déchet *m*	desperdicio *m*	rifiuti *mpl*
arboresence *f*	arborescencia *f*	arborescenza *f*

Nr.	English	Deutsch
468	**weathering**	Bewitterung *f*
469	**web**	Bahn *f*
470	**web guide**	Bahnführung
471	**weight feeding**	Gewichtsdosierung *f*
472	**weld line**	Bindenaht *f*
473	**weld line**	Schweißnaht *f*; Bindenaht *f*
474	**welding**	Schweißen *n*
475	**wet, to**	benetzen
476	**wind up, to**	wickeln; aufwickeln
477	**wood flour**	Holzmehl *n*
478	**yellowing**	Gilbung *f*

Français	Español	Italiano
exposer à l'intempérie *f*	exposición *f* a la intemperie	invecchiamento *m*
bande *f* continue	rollo *m*; tira *f* continua	nastro *m*
guide *f* de baude	guía *f* de rollo	guida *f* di nastro
alimentation *f* en poids	alimentación *f* por peso	alimentazione *f* ponderale
ligne de soudure	línea *f* de soldadura	linea *f* di saldatura
ligne *f* de soudure	unión *f* soldada	linea *f* di saldatura
soudage *m*	soldadura *f*	saldatura *f*
mouiller	mojar	bagnare
enrouler; embobiner	enrollar	avvolgere
farine *f* de bois	serrín *f* de madera	segatura *f*
jaunissement *m*	amarilleamiento *m*	ingiallimento

English

Deutsch

Français

Español

Italiano

Notizen

Notizen

Notizen